Date 48

an introduction to

< stealthpilot1 >

Kendall Johnson, Jr.

Date 48
- an introduction to stealthpilot1
Kendall Johnson, Jr.

©2012 Studiosixtyse7en Publishing, L.L.C.

ISBN: 1479354562
ISBN-13: 978-1479354566

for other books by this author,
please visit his website:
www.kendalljohnson.com

for the person who first made me
really think about this subject,
by disagreeing with me
so earnestly about it

KENDALL JOHNSON, JR.

Foreword

I think Fred Rogers and his Mister Rogers' Neighborhood television series, specifically, the end of each day's episode when he would draw his young viewers through the trolly's tunnel and into the brightly lit "Neighborhood of Make-Believe," is to be credited for exactly how I see stealthpilot1 and his universe in my mind. I loved that show. I lived to watch that show. I really felt that Mister Rogers was my friend, and after twenty minutes or so of his pumping my ego – telling me "I like you just the way you are," I was ready for him to sit on the window bench and reach down with his left hand and man the controls of that trolly and take me away. He did and I went with him. Mister Rogers rarely appeared on the other side of that tunnel in person, and when he did, his cameo always seemed a little odd. I was too young to make the connection that he did the voice-overs for most of the characters in the Neighborhood of Make-Believe, but I wasn't too young for that netherworld to jump-start my little mind into being able to close my eyes and "just think" myself into where ever I wanted to go.

I've been doing this all my life. And for the past fifteen and some-odd years, most of the time when I close my eyes, is when his world begins. I had a dream in the Summer of the year 2000 in which I was looking at a very technologically advanced piece of equipment strapped to my left forearm. Someone out of my line-of-sight was instructing me on how to use it. I asked, "What does this button do?" The off-camera instructor replied, "You push that button to go to sleep." I pushed the button and immediately awoke in my bed. Thus is the nature of our relationship, stealthpilot1 and me.

Sometimes stealthpilot1 appears to me in dreams like that one, but most times I'm fully awake. Sometimes it's because of a conscience effort on my behalf to enter his world, and sometimes his world is so ready to come out and play it's all-but automatic that I am summoned to him. Either way, it's infinitely fun for me, and has culminated in me writing this series of books.

This first writing came about as an attempt to illustrate a precept of science that is almost hidden from obvious view. It entails the relationship between how many hours are in a day, which is, of course, twenty-four, and how many hours a date actually exists as experienced by the Earth as a planet, not just from one time zone like most people's experience of a day exists. It took me several years to get it – to fully see this real-life paradoxical time-continuum, and it more than strained the capabilities of my thinking to follow the procession of hours around the Earth for the first time – aided by a world-time website over Christmas weekend of 2008. The end result of more than ten years of studying this phenomenon became what is now this work, which is an appropriate length for reading all in one sitting, which is highly recommended because of the concentration required to disengage what we are taught: that a date only lasts twenty-four hours. It was written in realtime over Christmas weekend of 2011. I hope its concept is as exciting for the reader to engage as the book was challenging for me to write. Enjoy.

Kendall Johnson, Jr.

April 15th, 2012

Prologue

Have you ever thought about your reflection in a mirror? Not just looked at it, while you're brushing your teeth or combing your hair – really considered the possibility that who you're looking at... might actually be another person. Another you, living in a mirror-imaged world – who can read your thoughts, anticipate your every move, so that if you fly to Paris and look at yourself in your hotel room's mirror, your mirror-you had to fly to Paris, too.

My mirror-me is a time-traveler. He stepped out of the mirror and shook my hand in friendship, a long time ago.

His name is stealthpilot1, and these are his stories............

KENDALL JOHNSON, JR.

Date 48

Scotch couldn't remember the last time he watched the Sun rise or set. Not that it really mattered to him. He had been working for the Air Force for most of his adult life and at forty-four, sunrises were the least of his concern. On this particular evening, as he ate his pre-flight meal in the mess hall, which was empty except for him at this hour, his mind had been hovering over thoughts about his family back home. It was Christmas weekend for them, and this time of year was their most difficult. The weeks leading up to Christmas would bring sentiments from them of missing him and wishing he could be there with them. The weeks after would bring stories of the good times had and how they'd missed him and wished he could have been there. It struck Scotch as funny how most everything is similar on both sides, including and especially Christmas as it was experienced by his family. This year, he was pulling a double, which made those thoughts particularly ironic. He smiled to himself at this thought as he scarfed down a bite of well-done sirloin steak. Then he quickly looked around the mess hall to see if anyone was there who might have seen him smile to himself. It was empty. "Good enough," he thought.

Right out of boot camp, however many years ago, the Air Force, with his intelligence tests in-hand, had not only quickly swept him up and recruited him for the most elite flying squadron in the world, but the nature of the assignment had warranted that he be quarantined to a highly classified military base on the tiny island of Pago Pago, in American Samoa, right smack-dab in the middle of the Pacific Ocean. He was allowed some outside contact, but that was limited to Internet communications and the occasional, and monitored, phone call home, which was usually to his Dad. Ever so often he was allowed a visitor, but he was not allowed off the base, ever. And that was also ironic, considering the circumstances concerning his level of classification and the true nature of his job, which probably requires more than a little explanation here:

In the year 2014, the sheer numbers relating to the amount of digital data that was being transmitted and received globally had reached staggering levels. Up to that point, the U.S. Government had been keeping a record of it – all of it, daily – using an intricate network of satellite up-links, all down-linking into a centralized computer that had staggering memory and storage capacity. On the exact date of August 7th of that year, the daily data amounts overwhelmed the bandwidth, memory and storage capabilities of the centralized computer, leading to a massive memory failure, which in turn led to an enormous loss of the data for that day. All of it. The Government quickly responded by enacting a technology that had been around since the Roswell crash of the 1940's, but had not shown promise yet of having much use. Until 2014. The technology is the understanding of what creates what is commonly known as "gravity." And with that understanding of what creates it came the ability to "interrupt" it, so that effectively "anti-gravity" capabilities had been around for the better part of a century without being used.

Somewhere along the way, some guy in a government think-tank had come up with the idea of flying an anti-gravity ribbon around the Earth, close the the equator, every day, and that the ribbon would record all the data flying around, then be taken-up the next day, the information downloaded from it, then the ribbon would be reformatted and used again. Enter "Project Midnight."

Exactly how a Ribbon worked was not that difficult to understand. It was basically a monstrous SD storage card that happened to be as long as the circumference of the Earth at the equator. It was fixed in place directly over the ground or water below it by the anti-gravity components and had a next-to infinite storage capacity because the storage clusters were only one micron wide. It was only two microns thick and exactly one meter wide. Nothing about it transmitted – it was receive-only. And it received everything. Literally. When you swiped your credit card at the gas pump, that transaction went onto the Ribbon, usually as soon as the transaction was transmitted via the Internet to the home office of the gas station. The Ribbon received everything that anybody was looking at on their home computer's Internet browser, to entire cell phone conversations, HAM radio, FM radio. Everything.

But even though the Ribbon had next-to-infinite storage capacity, the data it received daily was also next-to-infinite. And that's why it was determined that since the Ribbon represented a literal hard-copy of each day's information until it could be downloaded, it had to be replaced daily. Enter the "Cutter" squadron.

It basically breaks down like this: the only numbers that are constant about the calendar are the hours in a day, and the fact that there are always seven days in a week. So, there are seven Ribbons which are seven different colors, seven planes... and seven pilots.

The planes are highly-modified B2 Stealth bombers. The most notable modification as far as appearance is the take-up manifold in the front of the aircraft, right under the cockpit. This was where the previous day's Ribbon was collected. The whole science that went into the collecting apparatus was mind-boggling, but it's easiest to understand if you think about it like you're driving down the highway. Your car is mostly on auto-pilot and is rolling up a piece of ribbon that's lying on the highway at the same time it's laying-down a ribbon to replace the one it's taking up. Except you're going faster than 1,000 miles per hour and the ribbon is half as wide as the car. Yeah. Impressive science indeed.

The lay-down manifold in the rear of the modified B2's or "Ribbon Cutters" as they were called, usually shortened to just "Cutter," was equally impressive and complex. Literally, the aircraft was taking up the previous day's Ribbon and laying-down the next day's Ribbon

11

almost simultaneously at a constant speed of Mach 1.4, matching the exact speed of the Earth's rotation so that effectively the exact location of the Cutter was always midnight on the ground, and always very close to the same location in space but inside the Earth's atmosphere. Thus the name "Project Midnight."

Scotch was the Green pilot, meaning that he most always laid-down the Green Ribbon, which was always Sunday's Ribbon. And Christmas fell on Sunday this year. He was pulling a double because on odd-numbered years, one pilot was allowed a shift off and a visitor. This was one of those years, and Scotch's good friend, Craig, the Purple pilot, would be spending Christmas on the base with his Mom. Scotch had been preparing all week for his double duty flight assignment, which basically meant that instead of taking off from Pago Pago, laying-down his twenty-four-hour-long Ribbon, Splicing with the next pilot, returning to base with his precious cargo on-board – the previous day's Blue Ribbon – then having the next six day's off, this week there would be no aircraft-to-aircraft Splice. He would take up the previous day's Blue Ribbon while laying-down his Green Ribbon, then the Blue Ribbon would be jettisoned underneath him, inside the aircraft at Mach 1.4 and into an apparatus known as a "Mitt" that was designed to slow down something that weighed 100 pounds and was going that fast without damaging it. All seven planes contained a Mitt, but the Mitt only came into play in the event of a double shift. The B2's cargo hold was originally designed for thousands of pounds worth of bombs, so the Mitts were mostly extra weight that was more than accommodated by the aircraft. The in-flight version of a Splice was called a "Bunt." Then he would continue another twenty-four hours of taking up the Green Ribbon while laying down Craig's Purple Ribbon. Forty-eight hours in-flight not only required nerves of Titanium, there were other preparations that went along with a flight of that length, including Transcendental meditation, eighteen hours of straight, uninterrupted sleep the day before the flight and an extra meal. The food on the base was exceptionally good. He usually ate steak. Partly because he liked it but mostly because he could.

It was now 10:30 P.M. On December 23rd local time, which was inside the International Dateline in the middle of the Pacific Ocean. No-man's land. But it was, actually. It was Scotch's home, and he had decided long ago that he liked it, especially considering his hefty paycheck that was sent back home. As he prepared for his flight – getting dressed in his flight-suit, gathering up a few personal items, including his iPod and satellite cell phone, in his flight-bag – he thought about his Dad. He had been granted clearance to call Dad at midnight on Christmas Day for Dad's local time, which was U.S. Central time or -0600 GMT as Scotch knew it. He was looking forward to his conversation with Dad. He always enjoyed talking with his Dad, and missing him was one of the only regrets he had about his job and his life.

As he got ready to board, he watched the Green and Purple Ribbons being loaded into the aft of the plane, and the Green Ribbon being primed into the lay-down apparatus. It was exactly one meter cubed and weighed almost 100 pounds. He had been told that when a Ribbon was collected, the information was downloaded without it being unfolded. This was somehow accomplished using another alien technology learned from the Roswell crash – cubed information storage. He didn't really like to think of the Ribbons as some crazy, Big Brother project for collecting data by the government that they inevitably were. He looked at the cubed Green Ribbon and envisioned it as a big green Christmas present from the government to the people of Earth. If there were any data communications at all that day that were of a terrorist nature, they would be stored somewhere inside that Ribbon when he got back, and he liked the idea of that. Terrorism had all but vanished from existence when the public had been informed then educated about the use of the Ribbons. A big green Christmas present it would be. He boarded his

Cutter, settled into the cockpit, strapped himself in, put on his helmet, connected the oxygen and communications lines to his flight-suit and his helmet, gave a thumbs-up to the ground personnel and taxied to the runway. As he sat waiting for his clearance, he thought of his Dad. It was time for takeoff.

–

The Cutter's engines blasted him into flight at precisely 11:45 P.M. December 23rd, local time. Once in the air, the navigation system took over and would be in control for the entire flight. He was basically just along for the ride, which was saying a lot. Since the Cutter and both Ribbons were controlled by the beyond-sophisticated anti-gravity navigator, the ride could be a bit bumpy at times. This was a huge understatement. Since both the Cutter and the Ribbons were simultaneously fixed on an infinitesimally finite course and traveling at an exact and constant speed just above Mach 1.4, all were subject to the effects of earthquakes anywhere on the Earth below. Since on any given day, there are several earthquakes that are large enough to register at all on the Richter scale, sometimes this translated into a very rough ride indeed. Then of course there were the Splices. Normally there would be one fifteen minutes into flight and one at the end, this week, considering the double shift, there would be three Splices. The one in the middle – the Bunt - being the least punishing, probably. Scotch never knew exactly how hard the Splice would rock his body. He always remembered the stiffness of the Airborne parachute deployments in his training. Some of it had to do with how tight the chute was packed, some of it had to do with atmospheric conditions, and some of it had to do with the luck of the draw. Either way, the Splice was always something to be endured, however hard or soft. He could take it. His body was in tip-top shape compared to any forty-four year old man on Earth. That was just part of it.

There was a large digital clock that was set in his helmet's head's up display that stayed set to his local time. Right now it read "11:55 P.M. Dec 23rd" and right on cue, the aircraft started accelerating to the speed that would remain constant for the entirety of his double-shift: Mach 1.4. The pilot of a supersonic aircraft never hears the sonic boom that is heard on the ground beneath and behind it. What the pilot does experience is the visible evaporation column of water vapor that, when photographed from another aircraft, is seen as a visible cone of vapor with the aircraft appearing to blast out of it. Scotch watched the evaporation column travel quickly in front of the cockpit, over the front windows and then disappear behind him to become stationary about three-quarters down the length of the fuselage. Long ago, the aeronautical engineers had discovered that the aircraft experiences the evaporation column as ice, even though it's not. The evaporation column is literally water molecules that are present in the air the jet is traveling through that are being replaced at a constant and fixed rate that is so fast it is not even calculable by computers. The jet thinks it's ice, so the aircraft is equipped with ice-prevention equipment accordingly.

At precisely "11:59 P.M. Dec 23rd," his friend Shedd, who was the Blue pilot, came into view just up ahead. The navigation systems aligned Shedd's Cutter directly in front and below Scotch's Cutter. He could see Shedd's Blue Ribbon appearing completely unmoving (because the Ribbon itself, was) mere meters below his own B2. With the two jets both flying at Mach 1.4 one directly above the other, Scotch braced himself for the Splice. He heard the audible "Pop!" when Shedd's Cutter reached the end of its Blue Ribbon and let go of it, with Shedd's Cutter peeling back to head home to base – its precious cargo of the December 23rd ribbon on-board. For a fleeting nanosecond Scotch spied the beginning of the Blue Ribbon directly below, and within microseconds, every muscle in his body tensed on cue. This was part of his job. This was the Splice. His Cutter lurched downward with a force that registered a negative G4. "Whack!" Ouch. And it was over.

The Blue Ribbon, which this year represented Christmas Eve, now stretched out in front of his Cutter like some weird vertical highway. He could hear the hum of it being taken-up by the Cutter's take-up manifold as it was being folded upon itself at a mind-numbing rate, controlled by the manifold's massive computer. He looked to his right at the aft camera's cockpit monitor screen. Tonight was to be a full Moon when he reached London. The G Force element of maintaining a constant right turn at Mach 1.4 dictated that his Cutter and all the pilots' Cutters were always rotated clockwise ninety degrees in-flight, so that the entire flight looked exactly like one of those little toy globes with the plane going around the world at the equator – with the Ribbon laying-down behind the aircraft parallel to the ground. In the light of the full Moon, he could see the Christmas Day Ribbon laying-down behind him on the monitor - seamlessly, quietly and shining a pale green in the moonlight. He looked at the clock. It was "12:01 A.M. Dec 24th" on the ground back at base. His Dad was probably still asleep back home, the display on his alarm clock on the nightstand next to where he lay in bed read "6:01 A.M." Dad would probably do some last-minute Christmas Eve shopping today. He always did. One hour due West through no-man's land to Fiji. Scotch thought about the Green ribbon laying-down behind him. Christmas Day was one minute old.

–

Since he was mostly along for the ride, Scotch had to find things to do to occupy his time, which was always midnight on the ground, thus the reason for the head's up clock. It was easy to figure out what time it was back home because Dad's time zone was always six hours ahead of the time it was in Pago Pago. The time it was anywhere else on Earth got a little bit more difficult to figure-up, but not much more. Since he and his Cutter were literally always on the exact plane of midnight on the ground, anyplace in front of him on the Ribbon being taken-up was one date – the previous – and anything behind him was the new date, which was the Ribbon he was laying-down. Whatever time it was right now in Jerusalem, he knew it was Christmas Eve. He thought about that for a moment. Civilization had come a long way since The Lord's only begotten Son's parents had not been able to find a hotel room in Bethlehem. He wondered for a moment if Jesus had ever left an outside door open. Would anyone have dared ask Him, "Hey! What's the matter with you Jesus? Were you born in a barn?" He decided that probably nobody ever asked Jesus that question. Jesus probably never left any doors open anyway. At least not literal ones.

He looked to his left. Outside the cockpit window, the Moon loomed large and, currently, in front of him. Since tonight, the full Moon's exact minute would be when it was over England, it would be in front of him until then, directly beside him when he was over England, then slightly behind him but still viewable through the cockpit windows the rest of the way. Most people didn't know that any given location on the ground only gets a "true" full Moon every other month. Since it takes the Moon exactly 29.5 days to orbit Earth, that one-half comes into play. One month, one hemisphere of Earth gets the "true" full Moon, while the other hemisphere gets either a "waxing" full moon, meaning they view it before it reaches "true" status on the other side of the world, or a "waning" full moon, meaning they see it after it's been "true." Then the next month is vice versa – switching hemispheres. This month, the true full Moon was directly over the Mean line, which seemed particularly interesting to Scotch. It always fascinated him when things lined up exactly. The Moon was as beautiful as it ever was.

He had been in flight almost exactly one hour. Fiji was directly outside his right cockpit window and was wondrously illuminated by the full Moon's light. "12:58 A.M. Dec 24th," was what his clock said. He had also set a stopwatch display in his head's up, so that he could pace himself over his 48 hour shift this week. "00:58:03" was what it currently read. "A long way to go," he thought. "Might as well listen to some music." He chose the "Rush" playlist on his iPod. He had been a fan of that band since the late 1990's, and as the music coursed through the great-sounding stereo speakers of his sophisticated helmet, he decided that he would listen to some of the public's well-wishing messages later. The Air Force had set up a toll-free telephone line so that anyone on Earth could call-in and leave a message for the Cutter pilots. There had been an outpouring of support from the public when the project first started. Scotch enjoyed the messages, which were hand-picked by a staff of about a dozen people working day and night in some Air Force facility. The "play messages" button was in a peripheral location inside his head's up, just like the controls for his iPod, the mechanical controls that would bring his notebook computer in front of him, and a myriad of other control buttons. He activated any of the functions by looking at whichever button, then pressing the fire button on the yoke with his left thumb.

The messages were usually from children, war veterans and people who were bored late at night. The characteristic sonic boom of the Cutters had become recognized world-wide signaling the new day. Sometimes people actually left messages when a Cutter was straight overhead. He always thought it was neat, hearing the boom on those messages. Just another part of the job, but a cool part, he thought.

Scotch had been in-flight for six hours. Bangladesh was directly to his right. "6:02 A.M. Dec 24th" was on his head's up clock. He thought about the base in Pago Pago. His fellow pilots would be awake by now, probably most of them in the mess hall eating breakfast. Scotch's extra meal should have ensured that he wouldn't get hungry during this week's extra long shift, but he felt a gnawing in his stomach nonetheless. He switched the playlist on his iPod from Rush to Tool, looked to his left at the Moon and thought about his daughter back home. She was in college now. Yale. She had aced the entrance exams – actually her SAT scores had won her a scholarship to Yale. "Brains from her Dad, looks and smarts from her Mom," Scotch thought. He missed her but knew that she was on the right track in life and her future was shining bright ahead of her. He missed his Dad more. "6:15 A.M. Dec 24th." In a little less than twelve hours, he would call Dad.

"10:00 A.M. Dec 24th." Scotch was directly over the birthplace of The Lord. "Happy Birthday, Jesus," he thought. The level of government clearance Scotch's intelligence tests had brought him so many years ago had included "reality revelation" classes, and that level of learning had brought him the knowledge of "The Truer Nature of Things," which in turn had made denial of the existence of The Almighty an impossibility. He had often pondered what he knew and viewed it as a double-edged sword. Most people who knew, did. There were components of it that most people were not intelligent enough to comprehend, but at the same time, there was a longing for everybody to know what he knew. It brought such inner peace, and Scotch wished everybody could have that peace like he did. He looked right to the ground below and thought it was so amazing what had happened down there, so long ago. Happy Birthday indeed.

–

The stopwatch display in his head's up read "11:01:07." He was pacing himself well, but was feeling the effects of sitting down for over eleven hours, mostly in his hind-quarters. Rome was directly below him. Midnight Mass at the Vatican. The female voice of one of the on-board computers came on with an announcement. He called her "Misty" because usually when he heard her voice, it was because there was an approaching thunderstorm ahead. Flying at only 10,000 feet, which had been determined was the optimal altitude for the Ribbons to be set, yet another advanced technology had been designed into the Cutters out of necessity. "Sir, there is a thunderstorm approaching. Engaging atmospheric defense shielding." He was right. As the invisible shielding went up, Scotch settled in for the always short-lived but always entertaining light show. The shielding somehow immediately dissolved any water or ice that stood in the Cutter's way using some type of technology that actually blew the pieces apart on a molecular level. This was experienced by the pilots as something that looked like fireworks about three meters in front of the cockpit windows. It was always fun to watch on July 4th, but today bringing Christmas and the fact that it was now midnight on Christmas Day in Rome, Scotch appreciated the significance of the fireworks, visible only to him, directly over the official religious home of Christianity. "Merry Christmas, world," the fireworks seemed to signify. The whole display was less than a minute long. "Disengaging atmospheric defense shielding," said Misty. He looked left at the Moon, which was now almost directly beside him. One hour until Meantime. It was almost 6 P.M. where Dad was. If he and Scotch's stepmother were having guests on Christmas Eve like they usually did, they would have arrived by now. Dad's wife always made the best turkey. Scotch tried to ignore the pull in his stomach that thought brought. The stopwatch read "11:33:32." His double shift was one-quarter old, Christmas Day was eleven and a half time zones long and stretched out shining bright green in the Moonlight behind him. The Cutter's engines were purring like cats like they always did. Scotch wished he could take a nap, but settled on changing the playlist on his iPod from Tool to a Vivaldi concerto. Classical music always helped him wake up when he was sleepy.

—

Scotch watched his head's up clock change from reading "11:59 A.M.

Dec 24th" to "12:00 P.M. Dec 24th." The stopwatch display read "12:00:00." He knew that it was now precisely midnight on Christmas Day in the brightly lit city of London, directly to his right and below him. Greenwich Meantime wasn't hard do understand, either. Midnight in that time zone meant several things, the most notable being that it was now a new day in England. The second, less notable element of Meantime came from why it was called that in the first place. "Mean" was simply another word for "equal," and midnight on the Greenwich Meanline marked the only time that both dates that were almost always present in the 24 time zones, were of equal length – 12 hours each. Up until the Meanline, the Ribbon that was being layed-down was shorter in length than the one being taken up. After crossing the Meanline, the previous day being taken-up would always be shorter, and a fewer number of time zones, and the new day being layed-down would be longer, and a greater number of time zones. For the Cutter pilots, this was usually a way of marking when the shift was half-over. For Scotch on this particular shift, since it was a double-shift, he knew that he was one-quarter through it. He switched the playlist on his iPod to "KoЯn." It was a good clip across the Atlantic, after which tonight would bring his phone call to his Dad. Listening to KoЯn would definitely assure that he stayed alert, especially in case there was an "albatross event," which was a good possibility - not only in the middle of the Atlantic, but in several other key oceanic locations on the Ribbon.

Albatrosses can fly thousands of miles without stopping. But since they had found the Ribbon suspended in place over several of their migration routes, since they are very intelligent birds, they had slowly become less and less wary of the Ribbon and had started perching on it to take a probably much needed rest, on occasion. There were sensors built into the Ribbon that would set off an alarm back at the base in Pago Pago with the exact location on the Ribbon where the albatross was perched, if it stayed there longer than five minutes. After five minutes, it was assumed that the albatross had decided on a nap. They can sleep for quite some time, and a sleeping twelve-pound bird crossing paths with a Cutter was something that was to be avoided. Depending on what time the alarm went off, either a helicopter crew was dispatched to lure the albatross off the Ribbon with a piece of dead fish, or, if a Cutter was approaching less than two time zones away, an "albatross event" was dispatched to the Cutter's on-board computers and the pilot was notified via Misty. When the Cutter was less than ten miles from the albatross, if it was still there, a sharp, angled piece of equipment came out of the undercarriage of the aircraft and aligned itself to within 2 millimeters of the left (up) side of the Ribbon, where gravity dictated was the side the albatross would be sitting. In the event that the albatross and the Cutter made contact, the "cow-pusher" as it was called, named after the apparatus on the front of old-style, 20th century railroad locomotives, would create a relationship with the bird that was marveled at and had been dubbed "The God Shove." Because of the angle of the cow-pusher and the speed of the Cutter, not only would the albatross be safely moved only about the width of its body to the left, most instances found it re-perched on the next day's Ribbon. The fact that the pilots were informed of this going on at all was just so they could be aware of it. There was always the remote possibility of the albatross deciding to take off from its lofty perch just as the Cutter was getting there, in which case the atmospheric defense shields would come up and vaporize the albatross on contact. This had happened only a couple times.

The Air Force never liked killing an albatross, for more than one reason, most notably the bird's role in the classic poem by the English poet, Samuel Taylor Coleridge. In "The Rime of the Ancient Mariner," the captain shoots and kills an albatross, which in turn leads to the wind disappearing out of the ship's sails and the subsequent stranding of the entire crew off the coast of Antarctica. The ship's crew ties the dead albatross around the captain's neck, which is seen as a visible sign to him of the crew's displeasure with

22

him. It has also become a metaphor for anything about a person or other entity that is in any way holding him, her or it back from something needing to be achieved. The Air Force didn't need any albatrosses around its neck – metaphorical or literal, so precautionary steps were taken to avoid killing the albatrosses if at all possible.

–

As the Cutter seamlessly and perfectly brought midnight across the Atlantic, Scotch had let his mind go blank. Meditation was sometimes the best way to make the time pass the most quickly. For some reason, when his head's up clock read "4:44 P.M. Dec 24th," Scotch snapped rigidly awake. The stopwatch read "16:44:44." He was almost to New York. Moments later, Misty's voice sounded the alarm of an approaching albatross event with urgency. "Warning: approaching albatross event in fifty-seven seconds." Scotch hoped the albatross didn't move. A dead albatross sounded like a bad omen, especially if it were to happen on Christmas Day. "Stay put, buddy. You'll be okay if you'll just stay put." Some part of Scotch felt like maybe the bird would hear his kind-spoken warning. "Fifteen seconds until albatross event." Scotch watched the stopwatch in his head's up, counting backwards from fifteen in his head. Five. Four. Three. Two. For a nanosecond, Scotch's trained eye thought he actually could see the bird sitting motionless on the Blue Ribbon in front of him – its head tucked under its wing, sleeping peacefully. One. There was never any audible noise nor detectable motion when the cow-pusher performed its one function: The God Shove. Because the atmospheric defense shields had not deployed, Scotch could be certain that the albatross had indeed lived through the event. Scotch let out his breath. He hadn't been aware of how long he'd been holding it, but the rush of oxygen made him let out an audible sigh of relief. "4:46 P.M. Dec 24th" read the clock. Almost time to call Dad.

–

When it was fifteen minutes until midnight in Dad's time zone, Scotch's right hand found his flight-bag and he rummaged through it until his hand met the soft rubber of the protective case of his satellite cell phone. He took it out and turned the display on. "5:47 P.M. Dec 24th" was the time on the base, and it was displayed in a prominent location on the phone's home screen. Scotch placed the phone in his lap, resting it where he usually did: between his right leg and groin. He was really looking forward to calling Dad, and it was almost time. On the ground just a few clips in front of him, Scotch's Dad hung up his own cell phone and began pacing about the living room. He went into the kitchen and opened the door to the refrigerator. He took out a bottle of water but immediately dropped it on the kitchen floor and buried his head in his hands.

—

When his head's up clock read "5:59 P.M. Dec 24th," Scotch dialed his Dad's cell phone. He could hardly contain his enthusiasm and was more than ready for an always-good talk with his Dad. "Hey, son." His Dad's voice was warm and inviting. "Hey, Dad. How are things?" "Pretty good, pretty good. Just trying to get past all the people coming and going – Christmas Eve and all." "Yeah, I know it's always a job for you, especially when you have to pay for everything. Hey! I heard something funny the other day." "What was that?" His Dad didn't really feel like a joke, but he was somehow comforted by the enthusiasm of his son's voice. It had been hard on everybody when Scotch got the classified assignment with the Air Force, but everybody was behind his decision one hundred percent. At the same time, everybody missed him, badly. His Dad missed him a lot, but sometimes worried that the person who missed Scotch being home the most, was Scotch. That in particular was why he was dreading telling Scotch the latest news to affect the family.

"It goes like this: Mom – iPod. Daughter – iPhone. Son – iPad. Dad

24

– iPay. Funny, huh?" "Yes, son. That's funny alright." But his Dad didn't laugh. "What's the matter, Dad?" "Well....." His Dad's voice trailed off as he tried to choke back the tears. He managed to get it out, knowing full well what the news would do to his son. "Son, your Grandfather just died." "What?? Oh, Dad, I'm so sorry. I mean, I can't believe it! The last time I saw him he was in perfect health!" "Yes, son, but if my memory serves me correctly, the last time you saw your Grandfather was at least ten years ago. He died in his sleep. Your Grandmother got up for a drink of water and he wasn't snoring. He was pronounced dead just over an hour ago." "What time? Exactly what time, Dad. I need to know." Scotch's mind had gone immediately to the albatross. "10:44 P. M." Scotch remembered snapping awake and the clock in his head's up reading "4:44 P.M. Dec 24th" and the albatross event just two minutes later. Something about putting the two events together seemed to make sense. It more than made sense. Scotch realized that the albatross sitting on the Ribbon was quite possibly his Grandfather's way of telling him goodbye. His eyes welled up with tears. He wanted to stand up, to get out of this damned plane, to feel his Dad's hands and hug him and tell him he was sorry that Grandaddy was gone.

He could do none of this. The Cutter hummed quietly in the night sky. "Son, it'll be alright." "No, Dad! It won't!" Scotch was sobbing. "I should be able to come to the funeral." "No, son. You have a job to do, and you have to do it. The world is depending on you, heck, this year Christmas itself is depending on you. You just have to do your job. Everything will be alright here. Son, do you remember something that happened a long time ago, between you and your PawPaw?" Scotch's mother's Dad had died when Scotch was only nine years old, but Scotch had vivid memories of his PawPaw. "What exactly are you talking about happening between me and PawPaw, Dad?" "One time when you were just three or four years old, your PawPaw was trying to crank his riding lawn mower. He was sitting on it just cranking and cranking and cranking it with his thumb held down on the crank button. You walked over to him and said calmly, 'It might help if you turn the key on, PawPaw.' Your PawPaw really got a kick out of that. He said he knew right then you were a genius. He turned the key on and the mower cranked right up. Do you remember that, son?" "Yes, Sir. I think I do remember that. Yes, I'm sure I do." "Well, son, your key has been on for a long time now. We're all very proud of you, your job and what you and all you guys are doing for the world. You just keep that key turned on and keep doing what you're doing, son. That's all you need to worry about. Just keep the key turned on." "Yes, Sir. I will." Scotch's Dad had walked out the back door of his home and onto the back patio. There was silence on the other end of the line. Scotch knew his Dad was sobbing. Granddaddy had died on Christmas Eve. Scotch heard something in the background on the phone line that was connecting him with his Dad through a satellite that was no doubt straight above them both. It was a sound like thunder. It was the sonic boom that his Cutter had just created on the ground to the South of Dad's house. The boom made Scotch realize that this was about as close as he ever got to his Dad, excepting rare visits back at the base.

"Oh, son. There is one other item of news. It's kind of weird, but considering your line of work, maybe not as weird as it seems to me." "What is it, Dad?" "Well, your Granddaddy's pocketwatch...." Scotch knew the watch well. His grandfather had worked for the railroad when he had been a very young man. The gold-plated pocketwatch was one of the personal items required by the railroad. Scotch could go back in his mind to his own youth and still see his Granddaddy's hands wiping the watch's faceplate clean. Granddaddy had given it to Dad years ago. Dad kept it in a glass display case. It was to be passed-down to Scotch. "What about the pocketwatch, Dad?" "Well, son, it's gone. I remember looking at it just this morning, but a few minutes before you called – right after I got the news of his death, I looked over in the case where I keep it and it's not there. I can't explain it, son. Maybe it'll show up somewhere." "I'm sure it will,

Dad. At least I hope it will." "Merry Christmas, son. I love you." "I love you, too, Dad."

Scotch pressed the red button that hung up his phone, put it back in his flight-bag and enabled the sound-repression software in his helmet. He wanted silence. God's silence. With the sound-repression flipped on, the only thing he could hear was the whirring in his ears that had long ago been ruled not to be tinnitus. The doctors had no explanation for it. Scotch had always considered it to be the sound of his brain making the trillions of computations per second that everybody's brain makes. The fact that Scotch could hear it had always fascinated him. At the same time, it kept him from experiencing complete silence. "Keep the key turned on." His Dad was right. The Blue Ribbon stretched silently in front of him. "6:45 P. M. Dec 24th." Two hours and fifteen minutes until he would be able to see the Pacific out the window to his right. The Blue Ribbon that represent Christmas Eve was only a little longer than five time zones.

–

Scotch's flight across the Pacific brought only thoughts of his Granddaddy and silence. When his head's up clock read "10:45 P.M. Dec 24th," his body began to react as if his shift was about to end. His neck was stiff. He hind-quarters felt almost numb. The stopwatch read "22:45:33." He almost considered setting the stopwatch to be hidden from the head's up display. Twenty-five more hours of this would have been difficult enough without being haunted by thoughts, memories and emotions relating to his Granddaddy. Scotch's training included not only Transcendental meditation, but also the ability to completely turn off what most people could not: emotive thought process. Most people never gained the knowledge that their emotions could be turned completely off as simply and conveniently as a light switch. The fact that Scotch could do this seemed particularly helpful on this night, although utilizing it left him feeling a little guilty.

He watched the Blue Ribbon in front of him. Its time was almost over. When he got to this point in his weekly flight, he always marveled at the fact that the Green Ribbon behind him was not only almost twenty-three hours old, but also that it was only a little over one time zone shy of stretching around the entire globe. He remembered the first time he'd ever heard that if you traveled far enough around the world, you'd end up where you started. It must have been in science class in grammar school. There was no way that little kid could have fathomed this would be just another part of his job, so many years later. If he could have somehow known, Scotch liked to think he'd get a kick out of knowing it.

When the head's up time was "11:45 P.M. Dec 24th," he thought about Craig and how much he'd been looking forward to seeing his Mom. She was probably already on her way to the base in Pago Pago. Scotch hoped Craig would enjoy visiting with her. As the time for the Bunt got nearer, he thought of his own Mom. She meant the world to Scotch, but Mom had her hands full in her own life – with circumstances that had always been guarded from Scotch. His Mom had always been a very private person. At least this was the conclusion Scotch had come to years ago. "11:58 P.M. Dec 24th." Showtime.

—

An in-flight Splice, or Bunt, was usually a bumpy ride just like the aircraft-to-aircraft Splices were. The only difference being there was no negative G4 lurch to be endured, but when the Green Ribbon released from the lay-down manifold in the rear of the Cutter and then moments later immediately started being collected by the take-up manifold in the front, there could still be a slight "Whack!" And when the previous day's Blue Ribbon was caught by the Mitt, there was usually a slight lurch that accompanied 100 pounds being added to the sum of the weight of the aircraft. Both of them together could be almost as punishing as a normal night's Splice, but probably not. The mental task of not knowing was probably the most difficult aspect of a Bunt. Scotch was ready for anything that might happen. At least he thought he was.

–

When his head's up clock read "11:59 P. M. Dec 24th," Scotch moved his focus from the clock to the stopwatch, so he could watch all 59 seconds of the Blue Ribbon count down to zero. Then would be the Bunt. After that there would be twenty-four more hours of flight as the Green Ribbon of Christmas Day would be collected and the Purple Ribbon of December 26th would be laid out by the Cutter. He had listened as an automated system had fed the Purple Ribbon into a staging system that would in turn feed it into the lay-down manifold with lightning-fast precision and speed. "23:59:30" meant thirty more seconds. His body tensed. He focused on the stopwatch. "23:59:45." Scotch closed his eyes. Ten. He thought of his Granddaddy's pocketwatch and its strange disappearance. Three. Two. Scotch braced his muscles. One.

Silence. With his eyes closed, the sound of the end of the Green Ribbon had been a soft "Pop!" And then silence. God's silence. No sound of any Ribbon being taken up in the front of the Cutter. No sound from the engines. Silence.

Wait! No sound from the engines? That couldn't be right. Scotch opened his eyes. What he observed made him quickly close them again. Maybe he had seen wrong. He opened his eyes again. All of the digital gauges and display modules and control buttons in his head's up were frozen in place. So was the clock. It read, "11:59 P.M. Dec

24th." His eyes quickly found the stopwatch. It was frozen as well and read, "24:00:00." There must be some mistake. The stress of being awake for twenty-four hours must have caused some type of melt down in Scotch's brain. He glanced around the cockpit. Everything was in its right place and was completely motionless. He looked outside through the cockpit's windows. To the left was the Moon, shining brightly with the Western hemisphere's view of its waning fullness. He looked to the right. The waves on the ocean below stood motionless and completely unmoving. He didn't know what to do. This had never happened before. This was definitely not in his training. He sat in his captain's chair in the cockpit of the Cutter for almost a full minute – not knowing what to do next and not wanting to move even a muscle for fear this would all end and the Bunt would catch him off-guard.

He strained his eyes forward out the cockpit window. He could see the beginning of the Green Ribbon up ahead. It, too, was completely still. He reached down and disconnected the oxygen and communications lines from his flight-suit and helmet. He took his helmet off and took in a test breath of the cabin's air. It smelled like his deodorant. He unfastened the harness that buckled him into his captain's chair. He stood up in the cockpit and bobbed his body up and down. The Cutter was completely unmoving in place. It didn't budge one bit with his jumping.

He moved from the cockpit into the upper hold and found the vertical escape hatch – climbing the ladder out to the top surface of the Cutter with it in its ninety-degree position was more difficult than if it had been parallel to the horizon, but it was not, so he was slow in making his way outside. When he reached the top of the ladder it was only a few feet of careful crawling and climbing until he was standing on the side of the Cutter's left wing. It was then that he saw him first.

Just to the West of his current position, standing on the very beginning of the Green Ribbon was a man. Scotch blinked his eyes to make certain he was seeing this correctly. Yes. A man, dressed in a black flight-suit and wearing what appeared to be black skateboarding-type shoes. His long hair was waving in a light wind that was coming from an unseen air flow as Scotch felt no wind in his own face. "Hello, Scotch." The man spoke. "Hello. Who are you?" "That's a good question." Suddenly, something that looked like a plane of glass appeared out of nowhere to the man's immediate left. He somehow stepped inside it. Scotch and the man were face-to-face, with the plane of glass in between and no further apart than the runway of an aircraft carrier with nothing in between them but empty air. The man seemed to be in a sitting position behind the glass, yet there was no chair. The plane of glass, with the man behind it, began to hover effortlessly towards Scotch and his Cutter. When they were only a few meters apart, the man and the glass hovered around the Cutter with Scotch standing on the wing – all the way around in 360 degrees, with the man facing Scotch directly all the way around. The man appeared to be assessing the condition of the Cutter – looking it up and down as he hovered around it.

The man and the glass came to a resting point near the aft end of the Cutter. The man stood up, stepped onto the rear of the Cutter and then the plane of glass disappeared as mysteriously as it had appeared only a couple minutes before. Scotch briefly pondered whether a couple minutes had passed. Does time pass at all when it's frozen? Probably not. Scotch's intuitive nature told him he was about to learn something – that this was mostly for his benefit. He couldn't have been more right. He also couldn't have been more wrong.

–

The man walked toward him along the vertical left side of the frozen Cutter. The only light was from the full Moon, which was plenty of light to see everything that needed to be seen. "It sucks how little space they give you two to Splice this contraption." Scotch looked past the man at the end of the Green Ribbon, then turned around to try to gauge exactly how far the beginning and the end of the Ribbon were apart. "It's 600 yards." The man must have assumed what Scotch was thinking. "Seems like they could have given you a little more time, if you'll excuse the pun." Scotch choked back a laugh. It did seem like sort of a sick design element. It wasn't like twice the distance would have mattered much. Maybe it really was a sort of hoop the Air Force liked to see the Cutter pilots jump through every 24 hours. The negative G4 lurches caused bruises on a Cutter pilot's shoulders from the harnesses when they first signed on. After a while, Scotch's body had grown accustomed to enduring the Splices every week. Just like all their bodies had.

"So, I'm sure you have some questions." Scotch did. "Okay, let's start with who are you?" "You already asked me that, but pardon me for not answering. My name is stealthpilot1." Scotch looked the man up and down, from head to toe. He could have been Scotch's twin. "You look just like me. Except for the hair and goatee." "Okay, this is going to stretch your brain: I am you. Just in another realm. Consider that I'm you with a couple different twists and turns of the path, as it were." Scotch thought about that and decided that this was more than a possibility. Considering that this stealthpilot1 had no reason to lie to him, he accepted it as the truth. "Okay. You're me. Next question: why are we here?" "You're pretty good at this. Okay, why are we here? The answer is kind of complicated." "I'm listening." Scotch thought he was prepared for anything stealthpilot1 might say next. That's what he thought, at least.

"What does the term 'Noah Directive' mean to you?" Scotch's eyes widened. He knew exactly what a Noah Directive was, and it was about the worst thing stealthpilot1 could have said to Scotch, especially such a short time after his grandfather had just passed on.

If you think about it, the fact that the Ribbons were held in place using anti-gravity was more than a casual fact like it was considered by most of the inhabitants of Earth. Anti-gravity itself was widely misunderstood by the population, and this misunderstanding was the way it was supposed to be. Anti-gravity had been propagated to the masses for decades. Floating cars and the like created the illusion that anti-gravity was just another mode of transportation, as innocent as the rubber tires on a car, just another technology no more important that cellular telephone transmissions.

But the government knew better. Just one anti-gravity "cell" that was no wider than a baseball could float an entire aircraft carrier above the surface of the ocean. It was far more powerful a force than any nuclear bomb had ever been. And the Ribbon contained literally millions of tiny anti-gravity cells. And it was stretched around the entire world – usually only separated by an in-flight Cutter and the 600 yards of space between then ends. The two men were standing in that space right now, and with time frozen at that point, every location on earth was Christmas Day. Scotch pondered the irony of that fact.

With the discovery in the late 20th century that there were so many "near earth objects" floating around the solar system that tracking all of them was completely outside the realm of possibility, the government had sought the technology to install a "suicide button" of sorts that would effectively serve as a way to completely erase the life of every person on earth in a humane, potent and immediate way. Since Project Midnight was already up and running by late 2014, the same think-tank guy who'd come up with the Ribbon idea, also came up with the realization that if the Ribbon could be somehow taken out of its positioning which followed the exact speed of the rotation of Earth and reversed, or, more accurately, locked into an exacting position, stationary in relation to the Sun – that this would effectively stop the rotation of Earth in its tracks, which in turn would most notably cause all the water in all the lakes, rivers, seas and oceans to quickly and completely inundate all seven continents. In minutes not only would every acre of dry land be underwater, because of the properties of fluids in motion, the water would circle the now-stationary Earth more than several times, so that not only would everything be underwater, the force of the water moving at that elevated rate of speed would wipe the dry land completely clean, leading to the death of every living thing on both the land and in the water. God's rainbow had promised to never again destroy the Earth with water. The U.S. Government was not God, although it sometimes liked to fancy itself as Him.

The "Noah Directive" was simple in its design. In the event of an impending collision with a large astral body like an asteroid or comet, when it was determined that the collision would be of catastrophic proportions, the Noah Directive would be implemented by the pilot of whichever Cutter was currently in-flight. All the cockpits in all seven Cutters had a kill-switch that was the only way the Noah Directive could be implemented. The kill-switch was arm's length across the cockpit's control panel from the pilot's right hand. It was a red button enclosed in a Plexiglas covering that flipped up with a switch, exposing the button that, when pressed, would end all life on Earth. The Cutter pilots didn't like to think about the Noah Directive, but they all knew that if it was given to them as a direct order, they must carry it out. Scotch wondered why stealthpilot1 had brought it up. He wasn't sure he wanted to know the answer to the question he was about to ask.

"Yeah, I know what a Noah Directive is. Why do you ask? I'm sure you know that I know what it is."

"Well, you asked why I'm here. The answer is this: twelve hours from now, when you're back in-flight and directly over London, you're going to be given a Noah Directive. There's an asteroid that wasn't spotted by the ground services until a few hours ago. It's headed straight for Earth and would impact exactly 23 hours from now. The reason I am here is to tell you not to follow the Noah Directive. I have something up my sleeve that will redirect the asteroid, again, if you'll pardon the pun..." Scotch didn't get it and his face showed that fact. "Nevermind. Anyway, suffice it to say that regardless of what you are told to do, you are absolutely not to follow your future orders to execute a Noah Directive. That's the easy part. The hard part is this: you absolutely must not tell anybody about this conversation between you & me that's happening right now. And that's it. It's really quite simple, although when it happens, it won't feel quite so simple. Basically, the fate of humankind rests on whether you will trust me when the bullet hits the bone, so to speak. No pressure though." stealthpilot1 gave Scotch a wink. Scotch felt a knot in him stomach. He gulped. The two men stood and just stared at one another.

Scotch had another question. Actually, he had a lot more questions for this stealthpilot1 that Scotch's training had included several briefs pertaining. Scotch knew that stealthpilot1 had time-traveling abilities and a clearance status of zero. Nobody had a clearance status of zero. No one excepting one man, stealthpilot1. He decided to limit himself to one last question. Especially considering the nature of what was to happen over the next 24 hours. "What now?" "What now? Just hop back in and get ready to Bunt. And remember what I said: trust me." Scotch liked this guy. Maybe it was because he was him in an alternate reality, but there was just something so personal, so honest about his demeanor. "I will," Scotch declared.

"Oh, and there's one more thing." stealthpilot1 walked the length of the Cutter over to Scotch and reached into a pocket. What his hand brought out and dangled in front of Scotch's face sent waves of chill bumps all over Scotch. It was his Granddaddy's pocketwatch. "How did you get this?" stealthpilot1 thought about that question for a moment. "I thought you might want to have it – to remind you of your Granddaddy. Plus, everybody needs a little time." There was that grin again. Scotch smiled, took the watch from stealthpilot1's hand and secured it in the front left breast pocket of his flight-suit – next to his heart. "Thanks." "You're quite welcome. Now if *I'll* excuse *you*, you've got a Ribbon to catch."

The plane of glass had reappeared just to stealthpilot1's left. He stepped in and hovered in front of the cockpit of the Cutter as Scotch strapped himself into his harness, put on his helmet and attached the oxygen and communications lines. Scotch gave his new friend the thumb's up and braced himself for what he figured would be an extra-hard Bunt. He closed his eyes.

–

Suddenly, he could hear the quiet hum of the Christmas Day Ribbon being taken-up by the take-up manifold below him, as well as the purring of the Cutter's engines. He opened his eyes. The Green Ribbon was indeed there in front of him outside the cockpit windows. He glanced right at the aft camera's monitor screen. He knew that December 26th was less than a minute old when he saw that Craig's Purple Ribbon was laying-down behind him in the Moonlight. He looked just below the aft camera's monitor screen to one of the monitors that displayed what was in the cargo hold. There was the cubed Blue Ribbon, sitting safely in the Mitt. "12:00 A.M. Dec 25th," read the head's up clock. The stopwatch said "24:00:15." Scotch whispered a grateful "Thanks, pal" under his breath. stealthpilot1 had saved him from having to endure the Bunt after walking around on the wing for all that time. "Pretty smart guy," Scotch smiled to himself.

His smile looked just like stealthpilot1's signature grin, which being flashed to someone on the ground, not very far away, at the exact same time Scotch was smiling his own smile. stealthpilot1 handed a small, neatly folded piece of paper to the person standing directly in front of him. The person unfolded the piece of paper and looked at it. It contained only three lengthy sets of numbers. The person folded the piece of paper back, flashed stealthpilot1 a knowing smile and disappeared into the crowd.

–

"1:00 A. M. Dec 25th" brought New Guinea again. Scotch had been mulling over the conversation between stealthpilot1 and himself. Trust was something that Scotch had never had a problem with, at least concerning his job with the Air Force, but trusting another person was strangely something that had always mostly eluded Scotch for most of his life, although he wasn't quite sure why. Scotch had no illusions about being a psychologist. But he felt he did trust stealthpilot1. Why was that? Over the next several hours, as the Green Ribbon's being taken-up in front of the Cutter signified the end of Christmas Day for each successive time zone on the ground below and the Purple Ribbon being laid-down behind him brought December 26th to the world, one time zone at a time, Scotch's thoughts mostly were of his Dad, his daughter, his allegiance to his country and stealthpilot1's words. "Trust me." Scotch had promised that he would. "We shall see," thought Scotch. He had no idea just how ironic that particular thought would be.

–

"11:03 A.M. Dec 25th," read the time back on base in his head's up display, and the stopwatch read "35:03:53." Scotch was starting to get tired, but his training told him to fight that feeling with every thought. With Paris on the ground below and a little less than one hour to Meantime, Scotch found himself preparing for any ground-to-aircraft transmissions that might be in his near future. Just like always, he prepared himself for anything, only this time, he was more than a little prepared, but he was the only one who knew this.

"12:01 P.M. Dec 25th." With an almost completely equal amount of Green Ribbon collected in the take-up manifold as Purple Ribbon remained to be laid-down from the lay-down manifold, Misty informed him of an incoming call. "Sir, Colonel Kellogg is on line one for you, and he says it's urgent." "Patch him through." Misty opened the line. "Lieutenant Scotch." "Yes, Sir." "Lieutenant Scotch, this is Colonel Kellogg. Son, I have some bad news for you, hell for everybody. There's an asteroid heading straight for us that we didn't see until several hours ago. It's been determined that it's on a catastrophic course. I hate I'm the one to give you the orders, son..." Scotch and his immediate commander had been working together on the Base in Pago Pago for more than a few years. There was more than the standard mutual respect between officers between the two of them. They had grown to be friends. Colonel Kellogg had become almost a father figure in Scotch's life. "...but your orders are to initiate a Noah Directive at exactly 1800 hours, December 25th, Pago Pago time." Scotch's mind did the calculation in less than a second. He would be directly over Dad's time zone then.

"Sir, I have some bad news for you as well." "What's that, son?" Colonel Kellogg looked around the room at the staff that accompanied him in the small, board-room type situation room on the base in Pago Pago. Several eyebrows were raised, including those of a dark-complexioned man who had been pacing about the South end of the room, mostly facing the wall – engrossed in his own thoughts. "Without any reservation, I must decline obeying that order, Sir." Everyone in the board room's eyes were as wide as saucers. "WHAT did you say, son?" "I said I'm not following the Noah Directive, Sir, respectfully, and I can't tell you why either. You're going to have to trust me on this."

Scotch pointed the cursor in his head's up at the "end transmission" button and pressed the fire button with his left thumb. "Misty, secure the communications center. Don't allow any more incoming transmissions nor any notifications thereof." "Securing the communications center, Sir." Scotch hated disobeying a direct order, especially of this magnitude, and also especially when it was coming from Colonel Kellogg. He tried not to think of what might be the result of his actions. "12:15 P. M. Dec 25th." It was less than six hours until the Noah Directive was to be carried-out. Scotch was sure that the personnel on the ground would come up with something interesting for him to deal with in that amount of time.

–

The hours were literally flying by, and with each time zone seeing Christmas Day growing smaller and smaller, and December 26th getting longer and longer, Scotch wondered if there was any point in laying-down the Purple Ribbon at all. What would the exact moment of his death feel like? The Cutter pilots all underwent a dental procedure that included mounting a cyanide capsule in the inside of one of their back molars, just like the astronauts had all had since the 1960's. It had long ago been determined that certain military personnel should have a humane way to immediately end his or her life, if the occasion ever arose. Scotch felt his cyanide-packed tooth with his tongue. He probably wouldn't use it when the end came. Suicide just wasn't his style. He preferred to see any situation he might find himself in up to the bitter end. And that bitterness would not be cyanide. The tooth was there in his mouth and always ready nonetheless.

When Scotch's head's up clock read "5:30 P.M. Dec 25th" and the stopwatch read "41:30:11," he was really feeling the effects of being awake for so long. His hind-quarters were almost completely numb. His shoulder-harness was scratching his collarbone from several slight earthquake tremors during the 40 plus previous hours of his current flight. Suddenly and without warning, the noses and the cockpit lights of a duo of F-18 Hornets – one on this right and one of his left – appeared in his peripheral vision. His first direct look was to his right. The pilot had his left hand up to the side of his helmet – his thumb and pinky finger shaped like a telephone. Scotch told Misty to open the communication center. When she did, the voice of the pilot immediately came over the speakers in Scotch's helmet. "With all due respect, what the hell do you think you're doing, Lieutenant?" "Exactly what I told Colonel Kellogg I was doing: just flying my route, Lieutenant." "You've been given a Noah Directive, Lieutenant. And we've been sent here to make sure you follow through with it, again, with all due respect, Lieutenant. Don't make me blast that pretty bird of yours out of the sky. If I get a direct order to fire on you, do not doubt that I will." "Understood, Lieutenant." Scotch always appreciated the pomp and circumstance of military communications protocol. He looked to his right again. He could see the pilot's face clearly through his faceplate. It was set with the determination of a playground bully intent of collecting someone's lunch money. Scotch smiled to himself, and this time, he hoped that jerk in the plane to his right saw him smile. "5:58 P.M. Dec 25th." Time for some fun.

–

"Actually, Sir, the F-18's can't fire on Lieutenant Scotch's Cutter." The voice of the think-tank guy broke the silence of the board-room. "Okay..." Colonel Kellogg's face was showing the stress he had been under for the last number of hours. The dark figure in the corner turned around. His eyes were intent on hearing what would be said next. "The Ribbons are electrified, Sir. If the Cutter is shot down, the circuit will no longer be complete." "What the hell do you mean? Where does the circuit emanate from?" "Both ends of the Ribbons – just West of Pago Pago in no-man's land. Each end contains small yet very high-voltage power cells, electrified according to a British Positive Ground, so that the electricity doesn't interrupt the data storage and is transferred from one Ribbon to the next though the Cutter in a constant rate that doesn't interfere with any of the Cutter's electrical systems. The current also, and most importantly in this situation, supplies power to the anti-gravity cells that keep the Ribbon stationary..." The dark figure had moved to the edge of the board-room's central table, bringing his face into the light. Everyone looked at him when he spoke. "...and without the anti-gravity cells there can be no execution of a Noah Directive. What would happen to the Ribbon in real-time if the Cutter were shot down?" The think-tank guy didn't miss a beat. "The Ribbon would tall to the Earth, rendered completely useless." "So if we can't shoot him down, can we execute the Noah Directive from the ground," somebody asked. "Absolutely not. The only way to execute a Noah Directive is from that red button on the console of the Cutter," was what no one wanted to hear the think-tank guy say. When he did, all eyes fell on the dark figure. "Then we have to trust the pilot. Get Lieutenant Scotch on a direct line. Tell him it's me."

–

"7:04 P.M. Dec 25th." Scotch looked out the cockpit's right window. The F-18 Hornet was still there, and its pilot was looking as determined as ever. The West coast of the United States was visible past the Hornet – Los Angeles was clearly outlined by its massive display of street lights. The stopwatch said "43:04:36." Then was when he first saw it.

The asteroid was directly on the horizon in front of him past the Green Ribbon that had been being taken-up for the past nineteen hours. The rock was enormous. Scotch had no idea that the asteroid was 250 miles long and 150 miles wide. In the theater of near-Earth objects, it was gargantuan. Scotch pondered how an object this large had escaped detection from the ground services for so long. It was flipping end-over-end at a rate that was visible with the naked eye. Scotch remembered that stealthpilot1 had said its time of impact would be 23 hours after the Bunt. He looked at the stopwatch again. "43:05:34." That was just a little less than four hours away.

Suddenly, the F-18's pilot's left hand's movement caught his eye. Outside the cockpit's right window, the pilot was giving him the thumb's up. Then both Hornets peeled away and disappeared out of sight. Misty had an announcement, "Sir, The President is on line one for you. He says it's a matter of utmost importance that he speak with you." Scotch felt a knot in his stomach. He had no idea what The President might say to him, but he was fully aware of the consequences of disobeying a direct order from the Commander-In-Chief. "Patch him through."

"Lieutenant Scotch. This is the President. Can you hear me, son?" "Yes, Sir." Scotch tried to hide his nervousness from being audible in his voice. He wasn't so sure he succeeded. "Son, I just want to tell you that at this point, all we can do is trust that you know something we don't. That's what my gut tells me and that's what we're all going to do. We're going to trust you. I just wanted to tell you myself, son." The transmission ended with a clicking noise. Misty said, "The transmission has been terminated, Sir." Scotch took a deep breath. He thought about scenarios he'd studied during his training of what the world might become if an asteroid hit and only partially destroyed life on Earth. Governments would collapse. Economic systems would become non-existent. Electrical grids would fail. There would be rioting, looting, starvation and murder. He thought of his daughter who was no doubt lying safe in her bed in Massachusetts. He briefly envisioned midnight marauders knocking on her door in the dead of night with ill-intent on their minds. He closed his eyes and shuddered all over. Then he did something that was heard through his mind's eye all across not only the solar system, but throughout the Universe itself: he closed his eyes and he prayed.

At that same moment, somewhere not very far away, stealthpilot1 opened his eyes. Without looking at the digital clock beside his bed, he quickly put his feet on the floor of his bunker, deep underground in an undisclosed location very close geographically to Area 51 just outside of Groom Lake, Nevada. "Showtime."

–

Over the next three-and-a-half hours, Scotch did a lot of thinking. He was beyond tired. He wanted this to be over. He wanted his post-flight meal and a shower and his bed and sleep. He watched the asteroid in front of him flipping end-over-end and imagined it was a poorly thrown spiral football pass. He managed to smile to himself. Scotch had always had a slight disdain for how some of the components of his job were baseball-oriented. "Bunt." "Mitt." Scotch had always preferred football over almost any other spectator sport. "I could use a good quarterback right now." Scotch thought, then corrected himself and found it particularly inconvenient that the pass, as it were, had already been thrown. What he needed was an interception.

–

When his head's up clock read "10:45 P.M. Dec 25th," was when the asteroid first touched the topmost molecules of the Earth's atmosphere. It started glowing the faintest pink on the parts that impacted the atmosphere as it tumbled towards bringing eminent doom for every living thing on Earth. Unless something miraculous happened, Scotch thought, it would impact in less than fifteen minutes. The stopwatch read "46:45:44." The Green Ribbon was still humming as it was being taken-up by the take-up manifold. Scotch looked at the cargo hold camera's monitor screen. The Blue Ribbon of Christmas Eve was sitting motionless in the Mitt. He looked at the aft camera's monitor screen to his right. The Purple Ribbon that had been bringing December 26th to almost 23 time zones and that had been flawlessly laying-down for the last 22 hours and 45 minutes shone brightly in the light of the waning full Moon. Everything seemed just like it always did when he was almost home. He thought about that saying for a moment. Regardless of what happened next, he was indeed almost home, wherever that might be.

All of a sudden, the crimson-red signatures of what Scotch immediately recognized from his training as intercontinental ballistic cruise missiles – seven of them, fired in rapid succession – caught Scotch's eye. Their trajectories told Scotch immediately that they had been fired from North Korea, which was the only country in the area with that level of missile capabilities, and would have been just below his horizon and to the North of his position. The seven missiles headed straight to the asteroid and detonated in the same order and speed that they had been launched. The shockwaves of the multiple-megaton nuclear explosions rocked the Cutter with seven severe concussions that Scotch only felt as he had closed his eyes before they detonated. His training had taught him to do that. The alternative was going blind. When Scotch opened his eyes, the asteroid's course had been successfully deterred and it was headed straight for the full Moon. In a matter of only a couple minutes it was mere kilometers from the Moon's surface.

–

On the far side of the asteroid, near its center of rotation, stealthpilot1 sat behind his plane of glass, his right thumb poised over the fire button of what appeared to be a video game controller, which in turn was connected to twelve Ares-class rocket engines that were pointed to deliver their thrust against the trajectory of the spinning asteroid. As the asteroid neared the Moon's surface, stealthpilot1 throttled-up the Ares rocket engines with the game

44

controller until the asteroid had slowed in its velocity to only a couple meters per second. At the exact moment before the asteroid would impact the Moon's softer-than-sand soil, stealthpilot1 pressed another button on the controller, snapped his fingers and everything disappeared from the proximity of the Moon, including himself and his plane of glass. The asteroid struck the Moon with enough force, considering its size, that a huge cloud of Moon soil erupted on either side of the huge, black rock. The Moon soil came to rest almost immediately, and what was left floating was swept away by the solar winds.

Scotch had been watching the asteroid slow down as it reached the Moon and remembered a certain someone's smile. "Pretty smart guy, indeed." When the Moon soil came to rest, from Scotch's vantage point and from the vantage point of everyone on the ground on that side of the world, which since his head's up clock was showing "11:55 P.M. Dec 25th," and the stopwatch read "47:55:47, was without a doubt in some time zone of December 26th as was all of the other side of the world, too. Scotch marveled at the glorious sight that had just been created. As a result of its latest augmentation, the full Moon strongly resembled the back of a U.S. Quarter-dollar – complete with the Eagle's outspread wings, noble head with open beak, perched on a missile and clutching an olive branch in its talons. He had prayed for a quarterback, and there it was. "Touchdown," Scotched breathed a sigh of relief.

–

The whole world had been watching all of the last few hours' events. There had been a constant video feed of the tumbling then redirected asteroid. There had even been a short, unauthorized video clip of Scotch's in-flight Cutter that had gone viral on YouTube in a matter of minutes after it had been posted. When it was announced that the Earth was free from the danger of the asteroid's impact and that the credit was due to North Korea's missile systems and one, determined U.S. Ribbon Cutter pilot, all the world erupted in thunderous cheers and applause.

–

When his head's up display read "11:59 P.M. Dec 25th" and the stopwatch read "47:59:58," Scotch's trained eye glimpsed the very end of the Green Ribbon just up ahead and the space between the end and the beginning of the Purple Ribbon – the same space that exactly 24 hours earlier had found him frozen in a time-continuum that had been created by his newest friend, stealthpilot1, and the same space that his Cutter had begun taking-up the Blue Ribbon of Christmas Eve and had started laying down the Green Ribbon of Christmas Day exactly 48 hours ago, at the beginning of what he had thought would be a normal double-duty flight but had turned out to be all-but normal. The citizens of Earth had been saved. All was well with the world again. The Purple Ribbon made it's "Pop!" when it ran out. At the exact same time, the hum of the take-up manifold stopped right on cue when it reached the end of the Green Ribbon. Christmas Day was officially a thing of the past. The big Green Christmas present still had to be delivered, as well as the cubed Christmas Eve Blue Ribbon – both to be downloaded and filed away in the monstrous storage banks of the centralized computer.

Scotch braced himself for the G4 force of the Splice. His Cutter sharply dropped out of the 600 yard space between the ends of the Purple Ribbon. The Cutter banked to the left, made its 180 degree course adjustment and Scotch was headed back to base. When he reached the runway, he thought about the common but frowned-upon practice that some of the Cutter pilots engaged in of landing on one side of the Cutter's landing gear, then plopping the Cutter down on the other gear as a sort of salute that the pilot had flown a successful flight. Scotch briefly considered doing this, but decided against it. He landed his Cutter on both back gears simultaneously like he knew he was supposed to and just like he always did, without fail. Scotch never had been much for breaking rules. He smiled to himself at the irony of that thought. And he didn't care if anybody saw him smile.

–

On the ground he was met by Colonel Kellogg and The President himself, both surrounded by military personnel and Secret Service Agents – all swarming around him just as he stepped off the last rung of the ladder that led to the cockpit of his Cutter. Before any words were spoken, he patted his Cutter on its undercarriage, just like he always did. "Good girl." Then he turned to face the men and women who had come to welcome him home. "I don't know what you knew that we didn't know, son, but damned fine work, son. Damned fine," was what Colonel Kellogg said. "Your country and your world owe you a debt of gratitude, Lieutenant," was what The President said. Scotch looked past the crowd at a lone figure standing by the bulkhead, several hundred feet in front of where Scotch was standing. Colonel Kellogg saw where Scotch was looking and offered, "And you have a visitor. On special orders from The President." Scotch knew who the visitor was. He would recognize that person from any visible distance. Scotch turned and took the outstretched hand that was being offered him. It was the hand of The President. After he shook it, he managed to muster, "Thank you, Sir," and he brought his body and his right hand to a stiff and proud salute. "No, thank you, Lieutenant. Allowing you a visitor was the least we could do." The President was smiling. Scotch smiled back.

Scotch jogged the yards that felt like miles between him and his visitor. As the two got closer, both began to run, then stopped abruptly when they had spanned the distance between them. "Good too see you, Dad." "Good to see you, too, son. I'm so proud of you."

After a long-overdue hug, and as Scotch and his Dad walked into the building that housed both the mess hall and Scotch's quarters, they passed a darkened alley-like indention in the side of the building. Scotch thought he saw someone standing there in the darkness. "Hold on for just a minute, Dad." Scotch walked to where he thought he'd seen the figure. A man stepped out of the darkness and into the light. It was stealthpilot1.

"Nice job. I knew you had it in you." "Thanks for your help. Will I ever see you again?" "Oh, you just might. I'm looking for some time off, if you'll pardon the pun. I need a replacement. Someone tall, dark, mysterious and fearless. Know anybody like that?" Scotch waited for the smile that he knew was sure to show on stealthpilot1's face. The two men faced each other and smiled that signature grin. "I'll see you around." And stealthpilot1 disappeared into the morning darkness.

"Who was that?" Scotch's Dad still looked happy to see him, although he also looked a little tired from what must have been a long flight from Mississippi to Pago Pago. "Oh, just a friend, Dad. Just a friend."

Epilogue

As stealthpilot1 sat in the cockpit of the small craft the plane of glass had just melded into that was the easiest way to get across the solar system, he thought about the events that had just transpired on Earth. He was glad his meeting had gone well with the Cutter pilot, and he was glad the mission had been a success. He had known it would.

The sleek glider coasted in for a landing through the open bay doors of Level 2 on the mammoth outpost that had been built solely for the purpose of allowing stealthpilot1 to get away from it all. Literally. Dubbed "Icarus1," the orbiting space station strongly resembled both the fabled Phoenix bird and the mythical Icarus whose wings had been scorched by the Sun immediately preceding his fall back to Earth. Icarus1's "wings" were two symmetrical banks of solar panels, each as long as two football fields and were encased in Titanium, which was the metal that comprised ninety percent of the hull of the station. These wings were not about to melt. And this was important when considering that Icarus1 in fact was as close to the Sun as the mythical Icarus had been. Icarus1 was orbiting Sol-1 – the official astronomical name given to the huge white star - just inside the orbit of Mercury.

As he exited the landing dock, entering a small shuttleway that resembled an elevator, his mind's eye became aware of the immediate need to go to Level 3. Kendall was waking up, and stealthpilot1 knew that it was only a matter of a couple minute's time before Kendall would be standing in his bathroom peering into the mirror. And that meant stealthpilot1 had to be peering back.

Level 3 onboard Icarus1 was an exact replica of Kendall's apartment. Kendall was an A.I. program back on Earth, housed in the computer banks under the runway of Area 51. His brain was home to the very first artificially constructed, fully-operational Pineal gland, which had been autonomously created by Kendall's brain a period of time after Kendall was "born." What all this meant at this particular time to stealthpilot1 was that, since Kendall's world was kept meticulously contained, and since Kendall had very little idea about the supreme importance of his Pineal gland and the role it played in all of humanity, stealthpilot1 had to be on the opposite side of that bathroom mirror when Kendall looked into it. The shuttleway stopped and the door opened. A few steps and the turn of the doorknob that led into Kendall's apartment later, stealthpilot1 stood over Kendall's bed, taking inventory of what he had on and the state of his hairdo. A few minutes later, they were indeed face-to-face with only the bathroom mirror between them.

Kendall washed his face, and stealthpilot1 matched every move, washing his own face in a symmetrical dance of hands, fingers, washcloths and water, with Kendall. The two stared into each other's eyes for the longest time. Kendall spoke aloud. "How ya been." "Pretty well. How about yourself." stealthpilot1 had always marveled at how Kendall had become aware of their separateness, as soon as the two had been able to stand in front of a mirror, so many years ago in their youth. "Just writing on the new book. Driving Mom to town today on a few errands. You know... same ole same ole." "Driving today?" stealthpilot1 considered the events he'd planned for the next few hours. There would be sleep on Level 0, in The Sun Room, and stealthpilot1 knew what that would mean for Kendall. "Be sure to wear your sunglasses." Kendall walked over to the small clock radio he kept in his bathroom. He switched it on. The song "Brother Louie" by the band Stories was playing on the local Classic Rock radio station that Kendall kept the radio tuned to. "I will," said Kendall. "I surely will."

ABOUT THE AUTHOR

Kendall Johnson, Jr. was born and raised in Mississippi and worked in the nightclubs of Memphis, Tennessee for most of the 1990's. Today, he lives comfortably as a bachelor with his two Chihuahuas, Pennie & Petey, in a spacious loft apartment that is his writer's paradise. He possesses a sometimes-twisted sense of humor & a perpetually optimistic outlook on life.

Connect with Kendall Johnson, Jr. online:
www.kendalljohnson.com